When animals in the wild come within
close range of someone who sits or stands
in a hidden location waiting for them
to appear, the hunters of old called them
"visitors."

Generally I wait for the visitors or animals
to show up and I get to see something —
a few rabbits, sometimes a roe deer.
But when I don't have any visitors at all, I
simply make them up. I sit back and start
musing about the animals that may have
walked here in days gone by.

win
✳ 1993

Journey to the Ice Age

Mammoths and Other Animals of the Wild

Journey to the Ice Age

Mammoths and Other Animals of the Wild

By Rien Poortvliet

Translated from the Dutch by Karin H. Ford

Harry N. Abrams, Inc., Publishers

4

The type of landscape that allows you to imagine
yourself living centuries ago, that's what
I like! If at all possible, without any of man's
structures in sight or any human sounds to
be heard.
Tucked away somewhat uncomfortably (after all,
that can't be helped) in a ditch overgrown with
reeds, I leisurely observe how night falls.
Next to me sits my good friend Ezekiel. And
a stone's throw further down, sits Tok, also
with his dog.

Wisps of smoke from my pipe happily
serve to keep the mosquitoes away.
And if a flight of ducks is good
enough to show up every once in a
while, I couldn't be happier.

5

If I can find a satisfactory landscape,
I like to spend hours sitting on a high
seat on a ladder.
Lazily leaning back in my chair
and only barely keeping my
pipe going, I sometimes get the
opportunity to admire a doe with
her young from right nearby.
    And then I feel richer than a king!

6

In the company of an unsuspecting rabbit, I am likewise entertained for a considerable time.

7

Sometimes I sit waiting stiff and numb from the cold; at other times I lie sopping wet after a long crawl through dry heather. But it is always worth the effort.

8

Especially on a night with a full moon and snow!
Then it is so breathtakingly beautiful that it
doesn't matter whether or not there are visitors,
because there is more than enough to admire.

9

It is hard to understand how it is possible that so many animals in the wild can still be seen in our overcrowded little country.

Hundreds of thousands of waterfowl.

Great numbers of geese every autumn.

Rabbits have been
with us for a long
time as well.

Sometime in the 13th century, rabbits were released in our country. Such a hunting ground with mostly rabbits was called a warren.

Farmers who had nearby farms hated these gray nibblers like the plague, but woe was them if they dared to touch the little dears....

According to a hunting decree of 1502, a rabbit poacher was punished by having his right eye cut out. And repeat offenders had their other eye cut out as well.

If you lived near such a warren and you had a dog, you had to have part of one of the dog's forelegs cut off.

In the case of house cats, their ears had to be cut off; this would keep them from going into the sandy tunnels.

To limit the damage, they dug deep ditches along the warrens

← with the steep bank on the side of the farm lands. But this didn't solve the problem completely.

Poaching continued nevertheless, with the crossbow or with the help of ferrets. ↓

Poaching by means of snares was something people in the Netherlands learned from the Spanish conquerors.

We also have hares and wood doves here (the farmers certainly would like to see less of them).

15

Pheasants were brought by
the Romans, a long time
ago, and they are
still here.

And we also have partridges here.

And pine martens.

The number of foxes keeps increasing.

21

I can still remember clearly
how, time ago, everyone
was talking about the fact
that roe deer were sometimes
seen in the dunes of Ouddorp.
Today the local population
is estimated at 250.
    Roe deer in the wild are
to be found in all the provinces.

22

Some bucks have antlers that are hardly worth noticing while others boast a magnificent set.

23

We've got plenty of wild boars.

24

I still find
these visitors
very exciting.

27

Fallow deer have been with us
for a long time as well.
Prince Maurits (1567-1625) bought hundreds
of them in England and released them here.

28

But they had been here before that time. They were either brought here by the Romans or they spread all over Europe in the 13th century, after the Vikings brought them along.

We also have an even bigger type
of deer, the red deer, and a number of
these still exist in the game reserves.

We are actually
quite spoiled
here because, with
a little luck, we
can get to see all
the different kinds of
deer

32

or a hawk at work.

A woodcock could suddenly take
off right in front of us.

34

All in all, there is
a lot to be admired all around
us — if we are
open to it.

We certainly share our small country with lots of others, but fortunately many of these are animals— in fact, there are more than you might think—

36

and sometimes they are
closer than you think.

But the fact remains that some species of animals have gradually disappeared from our landscape. It is now about 20 years ago that, hidden in a tiny little hut made of cut peat, I was fortunate enough to observe the courtship display of the black grouse; there were no less than 15 of them.

It's no use looking for them now; they are simply not around anymore. They really couldn't get along with us any longer.

The seals I used to watch during
the crossing by ferry are
no longer there.

The last otter was
run over and killed
in September, 1988.
In my great-grandfather
Zachary's lifetime
(1823–1903), otters were
common animals.

Zachary sometimes saw flocks of storks flying above him. There were ravens as well, and sparrows were everywhere.

His father, my great-great-grandfather
Cornelis (1787-1855),
could tell us some tales
about beavers!

Oh yes,

41

and, what's even more
amazing, about wolves!
All kinds of tales...

It is July 30, 1810, at dusk. On the widow Janssen's land near the hamlet of Bussereindt, Mrs. Eggels and Petronella Peeters are busy sheaving. While Petronella is working, her three-year-old son Johnny is playing somewhere nearby. At least she thinks he is. But when she happens to look up, he is nowhere to be seen. When she starts calling and shouting, she sees a wolf slinking away in the distance.

The child is nowhere
to be found. The next day,
a number of people from the
village start out on a search.
In the afternoon, Frank Reyders,
18 years old, finds a child's
forearm. Shortly thereafter,
the little left foot, complete with
sock and shoe, is found by
17-year-old Matthew Weyers. Later on,
Harry Bongers (42 years old) and Peter Spee
(16 years old) find a few more pitiful remains.

44

On August 2, three days later, three-year-old Henri Peeters
is carried off by a wolf.
August 13. Bartholomew Dahmen, 8 years old,
is keeping watch over the cow and the goat
on a small pasture in the woods,
together with his 10-year-old stepsister.
Sibyl and Gerald Timmermans, the
12-year-old son of their neighbor.
All of a sudden, a wolf appears
from the woods and
grabs Bartholmew. All
that is ever found
of Bartholomew are
his intestines.

A hastily organized hunt
to find the wolf ends
in failure.
Although the wolf is surrounded in the
Rijtbroeker Woods, he manages to break through
at a point where men armed with nothing
but pitchforks and similar tools are posted;
and so he gets away.

"In the far-away region of Bieringen, near Helden," on the monday evening of August 2?, Maria is on her way home pushing "a wheelbarrow with fodder." Her little sister Judith walks next to her. Suddenly a wolf appears who grabs Judith and drags her away.

"With lanterns, people search for the hapless victim until deep into the night."

The next day they find the intestines.

John Joosten, 17 years old, lived to tell the tale. He managed to reach a house in the nick of time.

46

On September 15,
11-year-old Pierre
Biessen is killed by
a wolf in the surroundings
of Vlodrop. On September 25,
a wolf attacks two
beggar children and
drags one, a boy of
eight, into a nearby
wood.

All over the province
of Limburg, people are
living in terror wondering
where the beast will
strike next. They are
afraid to go to their fields
and they keep their
children home as much
as possible.

47

The stories that are told
become ever more terrifying.
According to some, it isn't
just one wolf—

but actually a whole pack of man-eaters!
Others again tell that it is indeed one wolf,
but a wolf the size of which no one has ever
seen! There were people who had seen the
incredibly huge tracks with their own eyes.

49

During the winter months that followed,
no wolves were seen or detected,
but the fear remained.
At the sight of tracks made by a large dog or the sound
of rustling in the woods, people became petrified.

On May 27, 1811, the wolf did strike again
and little Mary Jennissen, who was three-
and-a-half years old, was dragged off.
Her body was found near "Gnitjens Gericht," not
far from Bussereindt.

The next day Catherine Ramaekers was grabbed by a wolf in the presence of her parents.
But as people came running and started screaming loudly, the wolf dropped the child after dragging her for more than 150 feet.

Catherine was badly bitten, but she recovered fully.

Altogether, within a period of one year, eleven people had been killed and several had sustained injuries.

The last victim to die was Servatius Pluijm.
He was four-and-a half years old.

It is not known whether the people ever succeeded in catching the hated child killer. There is no further mention of wolves in the records of the Region of Nedermaas.

While we may find the idea of "wolves in the wild" rather exciting, our fore-fathers couldn't find anything appealing in the presence of wolves here in our country. Their reaction was, exterminate the bastards!

And that was in fact what they tried to do in every possible way. They used the trusty old method of luring them into pits.

They would sit and wait
near a spot that had been
baited.

They used snares,
traps and other devices,
and they would dig out
the pups. Everything
that could possibly be
done to wipe out the wolves.

And they organized hunts, of course. Anyone who wanted to could participate in these hunts and this contributed to the danger. Carried away by the excitement of the hunt, people sometimes fired at anything that moved.

During one such hunt at Beesel in 1685, Herman Heeskens "sustained a head injury when he was hit in the head by a bullet." The affair was hushed up through payment of 20 rix-dollars (25 dollars).

Sometimes there were massive wolf hunts. On April 30, 1593, an army of thousands of beaters (between the ages of 14 and 60) stands ready at the break of dawn for an enormous drive, spanning the entire province of Utrecht. The men are carrying provisions for three days.

57

As the sun rises, the line going from the town of Woerden to the inland sea called Zuiderzee moves in. Equipped with clubs, pitchforks, and maces, they move in beating their drums.
To the right, a line has been marked off from the Zuiderzee to the town of Amerongen.
In the western area, any wolves found must be driven toward the river Vecht while, in the east, they are driven toward the heights of the Amerongse Berg.

ZVYDER ZEE

Eemnes

Eem

Amersffoort

Vecht

Vtrecht

Woerden

Wijck

Amerongsche Berg

Lek

Millers who lived high up in their windmills had the task of "keeping careful watch so they could signal with clearly marked signs to the folk participating in the hunt which way the wolves were going."

And the folks on the hunt just walked on and on, hour after hour. Every bush was given a thorough beating and ever so often some creature

would dash out noisily, scaring them half to death.

These were mostly roe deer. Sometimes a herd of larger deer would suddenly dash off. Especially in the low-lying peat bogs to the east of the Vecht and in the nearby area of Gooiland, a lot of animals were flushed out by the long lines of drivers.

But they didn't even try to catch
any of the roe deer, red deer, or wild boars;
they were after the wolves.

As soon as the wolves showed up, men on horseback pursued them in order to drive them to the nets that had been set on the heights of the Amerongse Berg and at a spot along the Vecht River.

And once the wolves had run into the funnel-shaped trap of nets, the opening was closed off and the killing would start. They used things like pitchforks, clubs, and axes to do the job.

Any wolves that managed to evade the trap weren't scot-free yet. They had to cross either the Vecht or the Lek and, on both rivers, men waited in boats for the exhausted wolves.

After the hunt, the nets were taken back to the churches where they belonged: 600 feet to the church in Westbroek, and just as many other churches; 1200 feet belonged to the church of Vinkeveen.

And so ended the obligatory participation of the villagers, who then had to walk back to their villages.

In any case, a man could
say that he had
            participated
in a real hunt. Otherwise,
there wasn't much
hunting to be done for
an ordinary villager
despite the fact that
he could see, day
after day, how creatures
in the wild devoured his crops.
It was strictly forbidden for him
to touch any of the animals.
So you could hardly blame a man
for setting a snare every once in a while.

That way, he could vent his anger occasionally, and, for once, get back at them! First you had to pile up some crushed oats in one of the small pools and then you could place the duck snares (made out of horse hair because it floats).

And the next morning, all you had to do was pick them up!

It's easy enough to hide one of those smallish ducks under your shirt if necessary. But it's a lot harder to make off with larger game. And it's also more dangerous! So they mostly just threw rocks or something. And even that was strictly forbidden. But some people were so amazingly bold that they had the nerve to place deer traps on the tracks of red deer or roe deer!

When the narrow board in the center is depressed, the two outer ones jam against each other.

69

This way you would obtain
some game, but
you couldn't stop
to think of the
possibility of
getting caught!
Poachers were
often punished
by having their
eyes cut out.

70

It was possible (every once in a while) to buy game. At the Smokehouse and Provisions Store of the Nobility at Hunting Lodge in Old Dresden, the following items were offered for sale in 1669:

861 red deer
616 boars
646 hares
751 partridges
65 capercaillies

20 geese
4 swans
15 bears
74 wolves
15 lynx

170 foxes
55 badgers
17 beavers
27 otters
13 squirrels

So there was plenty of choice.

However, the right to hunt → was strictly reserved for the nobles.

Everything— the animals in the woods, the
fish in the water, the birds in the sky, really
belongs to them. All of this is owned by those who sit
high up on their horses and don't
think twice about riding straight
through the crops.

Not only are tenant farmers supposed to shut up and bear it, they often have to contribute their services. Sometimes they have to be drivers and, at other times, they have to stand guard all day in the cold to keep the game from going the wrong way.

73

AESOP

In between things, he must
have looked at his master's
dogs while thinking about
his own dog who did not have one
of those fancy collars and who
had to stumble about on
three legs.

Those dogs were used to hunt wild boars, preferably the solitary older male.

When it was clear from various signs—such as his tracks and the height of the marks on the tree against which he would rub himself—that it was really a large old male and when his lair was also still warm, the show could start and the dogs were let loose!

The dogs had to get the old boar to take a stand and that often cost one of them its life! In the Middle Ages, hunting the solitary boar (in French "sanglier" and in English "singular") was considered a military exercise where the wild boar represented the hardened, fearless enemy; he was black, ugly, and lived in mud and darkness.

76

The noble deer was hunted with more ceremony. First of all, a suitable deer had to be found; at least a ten-year-old hart and preferably an even bigger one. There was a search for such a stag before the day of the hunt itself.

The tracking was done with a dog that had been trained to follow tracks while remaining on the leash and without making a sound. They paid attention to signs, of course; for example, the hoofprints,

was it the print of a doe or a buck? If the tracks of the hind legs "caught up" with those of the forelegs, it was generally a skinny or young and agile animal. That was not what they were looking for.

What could be learned, for example, about the height of the antlers by looking at the foliage? When a gamekeeper thought that he had gotten close to the stag, he would climb a tree from time to time. The findings of several gamekeepers were then discussed with the master of the hunt over breakfast.

During such a session, they carefully studied the droppings they had collected.

77

Then the gamekeeper whose stag had been "chosen"
led the dogs to the place where the stag had been
tracked. Once everyone had taken their places, the signal
"let the dogs loose" was given. The pursuit could last for hours.

Strict rules were observed for all aspects of the hunt on the antlered animal, including the gutting,

when the dogs were solemnly rewarded.

Hunting hinds and fawns was done with less ceremony; the objective was to obtain venison. Three men on horseback, three men with bows, and a small dog. The archers walked alongside the horses so the deer would not see them.

And unobserved they would also take up positions behind trees.

If you did not have to shoot very far, the crossbow could be useful for fleeing game (as long as you aimed considerably in front of the animal),

but its main purpose was to shoot animals that were moving slowly or standing still. An accomplished archer could shoot accurately with a crossbow up to a distance of 650 feet.

They also had crossbows with a special shape that were used to shoot bags of gravel instead of arrows. Crouching alongside a fake cow, you could sneak up and try to shoot a partridge with it. There was no way they could shoot a flying duck at 43 miles an hour. It was better to leave that to others.

85

Falconry, practiced like an art for centuries, was a sport for the rich

in which women could participate
as well. Here we see Jacoba van
Beieren of Bavaria, who lived
in the first half of the
15th century.

I wish I could have been sitting on my high seat during those days to look at things that we can no longer observe today. Imagine sitting there and admiring a capercaillie while a wild cat appears in the clearing.

Still, I'm surprised that the capercaillie takes off as soon as he spots the cat. I would have expected him to be braver.

Just imagine that you are enjoying a beautiful summer evening sitting high up on your 15th century seat, and one of the last remaining family groups of aurochs ambles past!

An aurochs bull with a shoulder height of 6½ feet.
today's dairy cow, 4½ feet.

The last of the aurochs died in Jaktorow, Poland, in the year 1627.

Aurochs could be
found in our country
during the Ice Age
and long afterwards

and they lived in the
woods as well as
on the plains.

For a long time, it was assumed that a smaller species had existed in addition to the big aurochs.

But not too long ago, it was found that the difference in size was due to the different sizes of bulls and cows of the same species.

For centuries, the aurochs was a valuable prize of the hunt.

It makes no sense whatsoever to
play the hero and try to kill a big bull —
a young cow is less dangerous
and tastes much better.

The situation was the same with the European bison, who lived here during the same period as the aurochs; no one in his right mind would go after one of these tough, old, and dangerous beings.

And certainly not in August or September when the bulls would stay near the cows during the mating season— at that time they were highly dangerous!

When this hunter about
800 to 900 years ago sent
off his arrow after lots of
stalking, he probably
aimed at the animal on
the left.

The wisent or European bison,
who today still exists in the
wild in Poland, was common
throughout almost all of Europe
and was known to be a
coveted catch.

The lynx certainly appreciated a piece of the wisent as well; when he was feeling brave he would sometimes take a chance on attacking a weak or young animal.

But a roe deer or the calf of a red deer was really the largest prey he would take on; and he would preferably pounce at his prey from a tree.

And when
there was nothing
to pounce on, the lynx
always managed to run
into something as he
stalked about.

He was happy with just about
anything — from a mouse to
a roe deer...

Our forefathers must have looked at the lynx
with admiration, but he wasn't of much use
to them. They were of course happy
to use his skin, but they could make
much better use of an animal like the
large elk who lived about 1,000 years ago.

If you had been sitting on that high seat at that time, you might have seen this fellow walk by.

In Northern Europe,
the elk can still be found
even today.

I find it strange to think that, in places where there are now shopping centers, elks once left their tracks.

A large elk could have a
height of 6½ feet at the shoulder.

The way they strip leaves
off the twigs is by folding
their slightly protruding
upper lip over a twig and
pulling toward the tip.

With that awfully short neck, it's sometimes easier for them to go down on their knees when grazing.

In severe winters, they can resort to eating buds on the trees. ↓

Just to show you how much bigger an elk is than a deer or a roe.

These are the palmate antlers we recognize today but, a very long time ago, there were elks with these strangely shaped antlers.

The tools people used to
win their precious catch seem
rather unsophisticated to us.

Imagine
what it must
have been like to
stand there and hear
him approach...

and how excited the people must have been to have several hundred pounds of excellent fresh meat!

They used all possible ways
to try and procure themselves
a piece of venison;
hunting was done not only
with a bow and arrow and with
spears, but also with pits.

They weren't too concerned about the fact that, with the hunting methods of those days, there was a good chance that the female would be caught while her calf escaped. They could not afford to pay respect to fair-minded hunting techniques.

Meat in the pot!
That's all that
counted, whatever the
circumstances.

It didn't make much sense
to spend long hours searching
for an animal that had been hit;
generally lynx, wolves, or a bear
would have long since found the fallen prey.

Up until about the 12th century, Bruin
the bear also lived in our low-lying lands
and he ate everything, including meat.

Lazy bears would
sometimes also
steal cattle;
people weren't
too happy
about
that.

No matter how much it hurt to give up
such a catch, you would be much
better off not to insist.

An all-out effort would be made to catch Mr. Bruin. The pursuit with dogs might last two or three days.

Often an entire
hunting party was
required.

But there were also men who dared to confront a bear alone. They would wrap strips of cloth packed with wood chips around their left arm and hold it in front of the bear's wide open muzzle.

With the right hand they would then thrust the knife. What courage!

Emperor Maximilian of Austria is supposed to have killed a bear all by himself when he encountered the animal on a narrow mountain pass.

121

How dangerous was it
to coexist with such predators
as bears and other wild animals
that roamed freely?
It really wasn't all that bad. Of course,
if you accidentally got too close and the
female thought that you meant to harm
her cub, things didn't look too good
for you.

The difficulties with predators —
bears, lynx, and wolves — only
arose when people began to
keep cattle near their homes.

123

Those were the days when it would have been best to have one of those seats high up on a ladder. You would have gotten to see them all— capercaillies, wisents, bears, elks, and wolves. When you look around today, there is no indication that those animals really walked on this soil at one time. There are no droppings, no trees with signs of rubbing, no scratching places, no tracks, nothing. It is just too long ago.

But I can easily show you a clear sign of the people who lived here a long time ago, some 4,000 years back. Near our house, where we sometimes walk our dogs, there is this burial mound.

It may be difficult to tell from just looking at it

but this is an intact
burial mound from the
Stone Age.

And to make it even more exciting,
there is a burrow in the rear of the
mound where a fox goes in and out!

That fox is totally unaware of all the things that are to be found in its home.
(I must admit, though, that I don't know either what interesting items may lie under our house.)

About 4,000 years ago, a man was buried in that mound. They laid him down in a squatting position with his face toward the east.
For the journey, they provided him with a few necessities: an axe, a scraper, a few fishing hooks, jars with food and drink, a set of bow and arrows, and a leather bracelet.

↑
This is the protective
wristband. When the
arrow was released,
the bowstring would whip
back and could lash hard
against the wrist. That's
the reason for the bracelet.

With this simple weapon
a man could accurately
shoot a roe deer from a
distance of 150 to 200
feet.

Flint cores, from which arrow points could be made, among other things, were hacked out of the ground in Southern Limburg. The rough pieces of flint were sold by traveling peddlers.

They used this part of deer antlers to hack the stones loose.

Once you had become the owner of a piece of flint, you didn't immediately start banging away at it.

First you carefully examined the core from all sides to determine what its possibilities were and what kind of tool you wanted to make from it. You could make various utensils from one of these pieces:

celt

awl

small knife

another type of knife

small saw

points for arrow and lance

And then you struck the first blow.

129

In working flint, different types of material were used for striking

so that it could be tapped or struck in different ways.

boys were fascinated of course, that's understandable.

Chip after chip was tapped away so as to obtain a sharp edge.

1   2   3

Boys learned how
to shoot with bow
and arrow by
playing at it until
they became
proficient.

Though they didn't have valuable flint
points, they managed just fine with
ordinary wooden arrows
whose points had been
sharpened;
and many a farm dog could
confirm that.

131

Even with all that though, they still weren't hunters.

To learn the trade of hunting, the best thing to do was of course to go along often with dad.

That's how the fine points were passed along from father to son.

133

And little by
little you
would learn

134

lots of things you wouldn't just know without being shown.
Gutting too had to be explained in detail.
How else would you know that, for example,
a wild boar does have a gall bladder (and that
it must be removed at once) while a roe deer
does not have one at all.

135

You had to learn how to read tracks.

And you had to learn to listen, so you would know at what point during his mating song the capercaillie went completely deaf, so you could jump on him at that moment without being heard.

From your father you learned how to catch a fish and how to empty the bladder of a hare. And if your father didn't have time, grandpa was sure to know how it's done.

This is more or less what the village looked like where the people who built the burial mound lived. From spring through fall it wasn't so bad.

But once the winter had arrived, life was hard here. When night fell and the people sat inside shivering around the fire, wild animals stalked hungrily around the settlement.

Just like the wolves did here in our country,
well into the previous century.

I spent many an hour right near
that burial mound. And I have also,
during moonlit nights, often waited
for the fox near another hill (there
are actually nine all together).

All that time I never gave
a thought to those hills around me.
I didn't know then what I know
now, namely, that they are prehistoric
burial mounds and are therefoe to
be respected.

And what bothers me a little
now is not that I didn't know
anything about it, but that I
didn't "feel" anything either.

Then I think of those burial mounds
that are even older, the dolmens.
How did they ever manage to get those huge
stones to the chosen spot? Vicar Johan
Picardt (1600-1670) was convinced that
the work was done by "terrible, barbaric,
and cruel Giants."

But that's not the case.
Let's imagine the province
of Drente some 6,000 years ago. A group of
farmers are dragging an enormous stone for which they
have big plans.

The stone, weighing many thousands
of pounds and tied onto a kind of sled,
is dragged to its destination with
the aid of wooden rollers. They do
this kind of work preferably in
winter when the ground is hard.
This is the way they would move
↙ the stone into the hole that had
been dug.

The hardest job was putting the capstones in place. →
To do that, they built up a ramp against
the standing stones. Afterwards they
removed the earthen
ramp again and the hill was held together with a circle of stones as in a wreath.

Onlookers probably
observed the army of
laboring men from a
safe distance.

An animal that was very
common in Europe at the
time of the builders of
dolmens was the wild horse.

In summer the animal's
hair was short.
In winter, when his coat
was heavy, he looked
more typical of the
shaggy prehistoric
horse.

After a snowfall, they looked like sugar-sprinkled doughnuts.

When their winter coats began to
shed at the beginning of spring,
they would groom each other.

In those days as well,
nesting birds would gratefully
use the discarded wool.

Wild horses were tough, sturdily built animals who knew how to defend themselves against their natural enemies, wolves and bears.

Stallions use their forelegs for kicking while mares kick backwards.

151

The horses
were a coveted prize for
people as well.
But it was so hard to
152 catch those fast animals.

The people who lived near Solutré, to the north of Lyon in France, had an easier job of it. They simply drove the herd toward the cliff. The enormous quantity of bones that can still be found there is estimated to have come from about 50,000 horses!

In the regions near the Netherlands, there were no such possibilities. There they tried to drive the animals into the marshes or to hit or grab one of them by sprinting, with the hunters running in relays.

154

They also used pits to catch the horses. With this method, they sometimes managed to get a totally unharmed animal.

At some point, this must have sparked an idea in some dreamer's head, namely, that you should be able to do more with such an animal than just use it for food.

What he planned to do didn't come easy of course, but eventually the triumphal day arrived!

Dogs became pets more or less in the same manner.

At first, people just hated wolves with a passion. Sometimes, especially in winter, they would become so bold that the people had to keep them away by throwing rocks at them. And always that melancholy howling all around you....

And people also got very upset every time
it became obvious that the wolves could
follow a track of blood so much better and
faster than the hunters, whose progress
was often slow and laborious.

It must have been children who, at some point, took in a wolf pup they had found. The adults were probably not too happy about it, until they realized that it was a nice, safe feeling having such a vigilant animal around.

And he soon became indispensable to the hunt.

And so we became friends.
And today I sit tucked away
in a ditch with
Ezekiel's woolly
head next
to me.

About 10,000 years ago, you could hear everywhere in the Netherlands the peculiar clicking sound that reindeer make with their large hooves when they walk

A side remark: With this observation about 10,000 years ago I am crossing a boundary and getting myself into trouble. Let me explain. My good friend Keesje Bok from Ouddorp, as well as other good friends for whom the Bible is the ultimate authority, firmly believes that the Almighty created everything approximately 9,000 years ago (not 10,000).

Scientists on the other hand talk about millions of years. I myself don't know. I personally feel that 9,000 years is much too short but, to tell you the truth, it doesn't bother me that much. I observe in admiration all the things that the Creator, blessed be His Name, brought into being out of nothing and I don't really worry about the fact that I have no idea when exactly that happened. Let's tolerate each other's opinions.

If you wanted to go along
with a party of reindeer hunters
in those days, you had to put on this disguise.

161

And then it was a matter of trying to get close to the herd of reindeer. Javelins were equipped with points made of stone or bone and,

← with a spear thrower, they could be hurled with a lot more power than was possible with the hand only.

They used the principle of leverage

These spear throwers were rather easy to make from antlers and they were often decorated with carvings.

This well-known spear thrower is pictured in various books about the Ice Age where it is known as "the defecating young ibex." But the longer I observe it, the more I think that it represents a hind giving birth. Another reason is that I know of a similar kind of spear thrower where the animal sits with all four legs folded under. And I have observed roe deer and hinds in exactly this position while giving birth.

163

Our forefathers did not mind that reindeer antlers often had strangely tangled shapes.

The people didn't use them for hanging above the fire place. The antlers were used to make handy items such as awls, needles, knives, spear points, fish hooks, picks, and digging sticks.

Reindeer provided people with 99 percent of the game they consumed.

80 cm.

Since people were hardened
against the cold in those
days, they did not require
much in the way of
clothing.

They generally walked
around scantily clad;
but when the winter cold
arrived, they donned their
furs. After all they had
lots of hides.

First the remains of the meat were
removed from the skin. Then the brains
of the animal (the quantity was just
right, as if made for this purpose)
were rubbed vigorously into the skin
in order to tan it. Finally it was
stretched around a kind of wigwam
and smoked, which kept
it supple.

Reindeer skins
were used to great
advantage as clothing,
roofing, and
bedding.

167

And you didn't have to worry about running out — there were plenty of reindeer.

But, let's be honest, it only becomes truly exciting when you get to see animals that have long been extinct. Animals such as elk, wolves, lynx, and wisent are still with us, even if they live rather far away.

But your pulse really starts racing when you are sitting there and you spot a giant deer — you can hardly believe your eyes! This deer that comes walking by has enormous tabletops on its head. These antlers had a span of 10 feet and could even reach 13!

In order to witness something like that you would have to go back in time about 11,000 years.

But then you would really see a sight worth seeing! Here's an animal that stood between 6½ and 8 feet high at the shoulder (I have drawn the man there just to give you an idea of the size of the giant deer). Since numerous bones of this deer have been found especially in Ireland, it is often called the "Irish deer."

171

Imagine having to walk around with a weight of more than 90 pounds on your head; and from time to time you also had to take a quick look around!

And what about suddenly having to lift that heavy head while foraging...

At least he is lucky that those huge shovels were placed in a slanted position, so he doesn't have the added burden of a load of snow.

But the hardest part was of course shedding the antlers; at some point, he would have 45 pounds on one side!

173

With antlers like these it was impossible to seek cover in the forest and these animals had to remain on the plains at all times.

174

This is how the antlers may have developed: from the spikes of a young animal to the complete set of an adult.

All in all, the antlers were a rather
cumbersome thing. The reason why this
large deer became extinct
is in fact often attributed
to its impossibly large
antlers; it was simply
a little exaggerated.

If he was good and fast in using them, he could deliver some impressive blows with those shovels. But for everyday use, it was simply too heavy a burden.

11.5'!

177

ICE.
At times a layer of hundreds of feet.

If, just for fun, you go back some 15,000 years, you end up in the last period of the Ice Age. The weather is cold and bleak. A strong fierce wind blows almost constantly over the tundra. The large ice cap had absorbed so much water that the North Sea is dry land. If you should want to, you could simply walk to England.

If you just keep on walking westward along the banks of the river Rhine, you automatically end up in London!

But sit down for a while when you are halfway there and look around to find out what kind of game can be seen.

179

Some strange animals walk
around in large herds.
← These are saiga antelopes
in their winter coats.
This is how they look in
summer.
These hairy hunks
are musk oxen.

Somewhere in the Netherlands during the Ice Age.

Where today there is maybe a
large residential area, with
traffic lights constantly alternating
their colors and all kinds of
things move busily about—

the woolly rhinoceros
lived long ago.

These heads certainly don't look a bit like they belong in Holland.

To ward off the unpleasant cold, the animals wore woolly pelts.

The woolly rhinoceros was somewhat smaller than the present-day white rhinoceros of Africa, but it did reach a height of 5.75 feet at the shoulder. The horn could be close to 5 feet long.

Since their heads move constantly back and forth as they graze, the front horn is sharpened through contact with the rough vegetation.

Hunting these animals was dangerous work. The rhinoceros is rather short-tempered and can attack with lightning speed.

A safer way to subdue the rhinoceros
was by trapping it in a pit.
What the men wanted was the meat
and the skin;
they weren't interested
in bravery.

It was sometimes hard enough just to store the catch; cave bears and hyenas, for example, often wanted to have their share — or the whole thing.

In those days there were
cave lions here as well.

And here they are, finally, the mammoths
(it was about time)! That shows you that you
should never come down from your high seat too early!

A scene in the Netherlands about 15,000 years ago—
You won't hear one of those idiots boosting his moped
and there is no poison in the ground anywhere either.

Imagine seeing one of these walk by! Somewhat different from a rabbit way below your high seat.

191

Only after they have gone by, you notice that the animals move noiselessly. The reason is that they have pads on the soles of their feet.

Also noticeable are the puffs of steam caused by their breathing, which show up at an unexpected level because these animals have such an unusual nose.

The tip of the mammoth's trunk has a different shape from that of today's elephants.

mammoth

African elephant

Indian elephant

People have been frightened out of their wits when, after a flood or landslide, they suddenly come across one of these giant skulls! →
Those holes in the front do look like eye sockets, but they actually constitute the nasal cavity.

Maybe mammoths use those puffs
of steam from their breathing
to make frozen tussocks somewhat
more manageable, who knows.

They pick up their food,
consisting of grasses and leaves,
with their trunks and then
stuff it in their mouths.

Mammoths use their trunk for drinking.

They smell with it.

And they can examine things very carefully with it.

They greet each other by lightly touching trunks.

You can pick up
all kinds of things with it
and cart them away.

And when you got tired of that
trunk swinging back and forth
all the time, you could just drape
it over a tusk.

196

And those they certainly had!
Tusks that is. Both males and
females had them and they
could make things very
clear with them.

The tusks of an old bull could grow to a length of 6½ to 7 feet! A set like that would weigh more than 450 pounds.

The tusks were very handy for disentangling small shrubs that had become intertwined.

Swinging his oversize teeth from one side to the other, he sweeps the frozen snow away in order to get to the grass underneath.

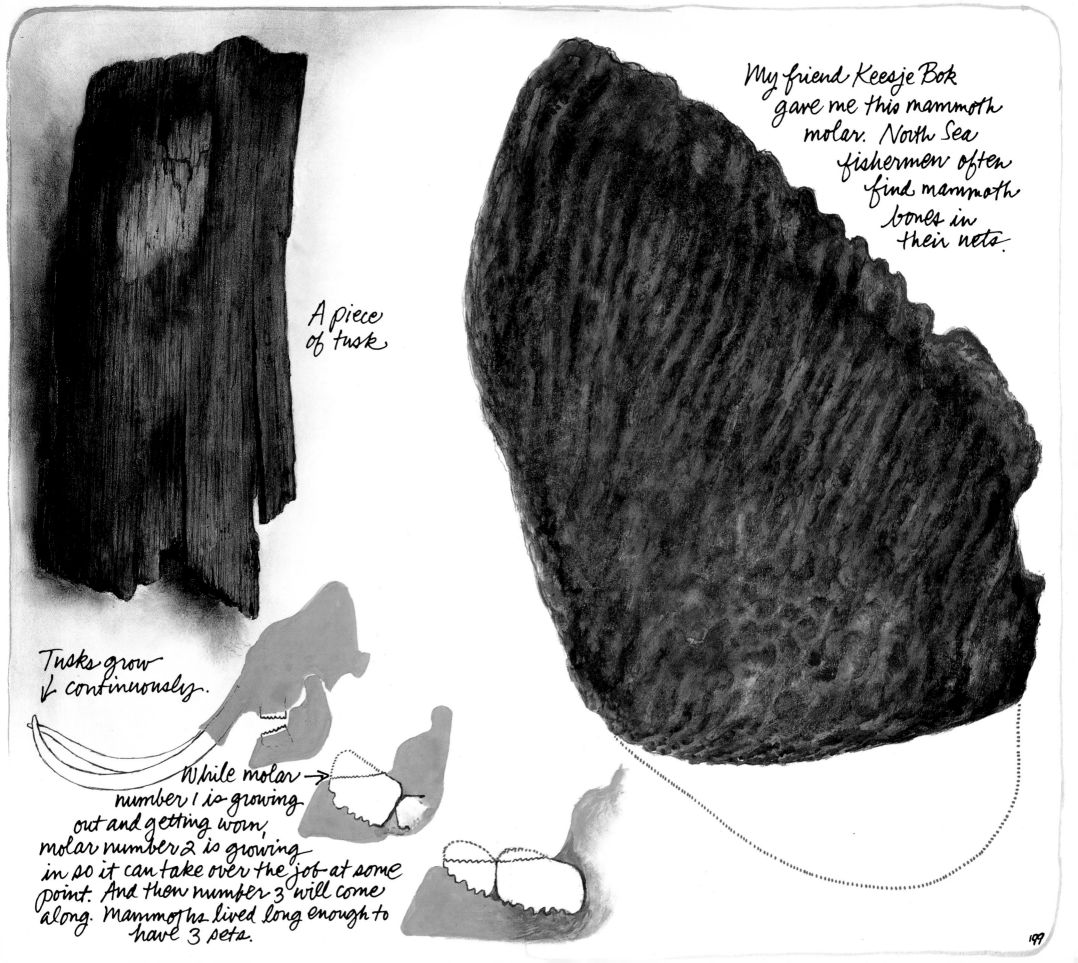

A piece
of tusk

My friend Keesje Bok
gave me this mammoth
molar. North Sea
fishermen often
find mammoth
bones in
their nets.

Tusks grow
↓ continuously.

While molar →
number 1 is growing
out and getting worn,
molar number 2 is growing
in so it can take over the job at some
point. And then number 3 will come
along. Mammoths lived long enough to
have 3 sets.

199

And look what I
have here,
real mammoth hair!
I didn't just draw this, but a
lithograph was made from the
actual hairs. The long hairs
come from the outer coat; the ones
that hang down from the shoulders
can grow up to a length of over
3 feet.

I could have simply picked up
those hairs near the tree where this
mammoth stood rubbing his big
behind! I don't know how you
feel about it, but I actually feel
somewhat moved when I sit here
looking at this tuft of hair.

This woolly tuft is
from the felt-like
undercoat. When you tap
it a little, tiny particles
fall out of it; they aren't
anything of importance
of course, but they are
about 15,000 years old!

It wouldn't surprise
me at all if mammoths
had heavier coats in
winter than in summer.

(Just as wild boars
look very different in
winter from the way
they do in summer).

Despite the bitter cold during the Ice Age, there was
a period during the summer when the top layer of
soil thawed out and all kinds of vegetation appeared.

During that period, the mammoths lived on the tundra
(that is, right here where we live today) where they
ate all kinds of grasses. When winter came, they
migrated to woodier regions.

Then they lived mostly on leaves, fir needles, and
bark, which they found along the banks of brooks
and rivers, where there were shrubs and trees.

Cold, cold, and still more cold.
Entire regions of the world were covered with a layer
of ice, which was easily several hundred feet thick.
And although they were well equipped to withstand the
cold, the mammoths often had a rough time of it.

For animals that weren't completely healthy, it was sometimes too hard a task to search out the 330 pounds of daily food intake.

From time to time, this old bull stops to rest his top-heavy head. Predators start to pay more and more attention to his way of walking.

Things certainly looked quite different
for him some 25 years earlier;
at the time, he was quick to show
who was giving the orders!
The younger bulls sometimes
didn't know where to hide.

And he made it quite clear to the older bulls who was boss and who, therefore, would claim all the cows when they were in heat.

The cows preferred to be serviced
by the largest and
bossiest bull.

At the end of a 22-month gestation period, the calf is born.

A mammoth cow has two breasts between her forelegs.

And when you observe her as she walks by, you can see them clearly.

Mammoths most probably lived to the age of 50 or 60.

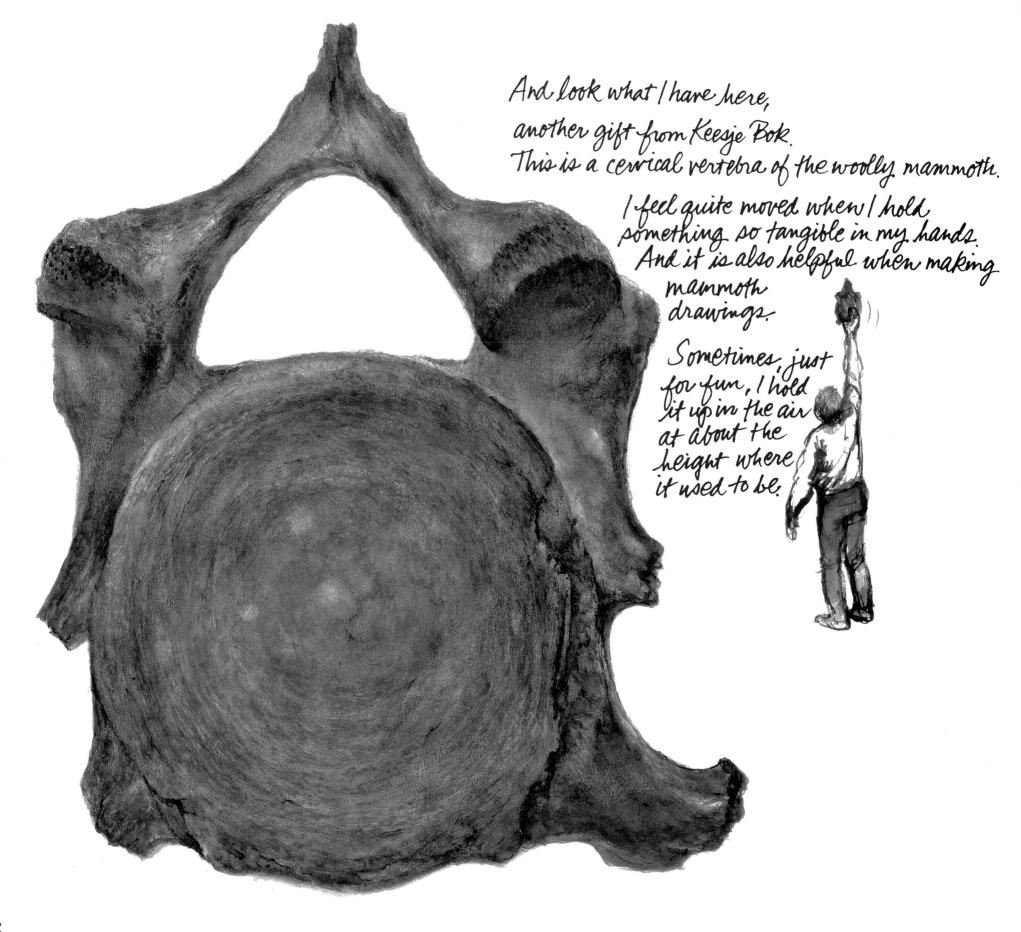

And look what I have here,
another gift from Keesje Bok.
This is a cervical vertebra of the woolly mammoth.

I feel quite moved when I hold
something so tangible in my hands.
And it is also helpful when making
mammoth
drawings.

Sometimes, just
for fun, I hold
it up in the air
at about the
height where
it used to be.

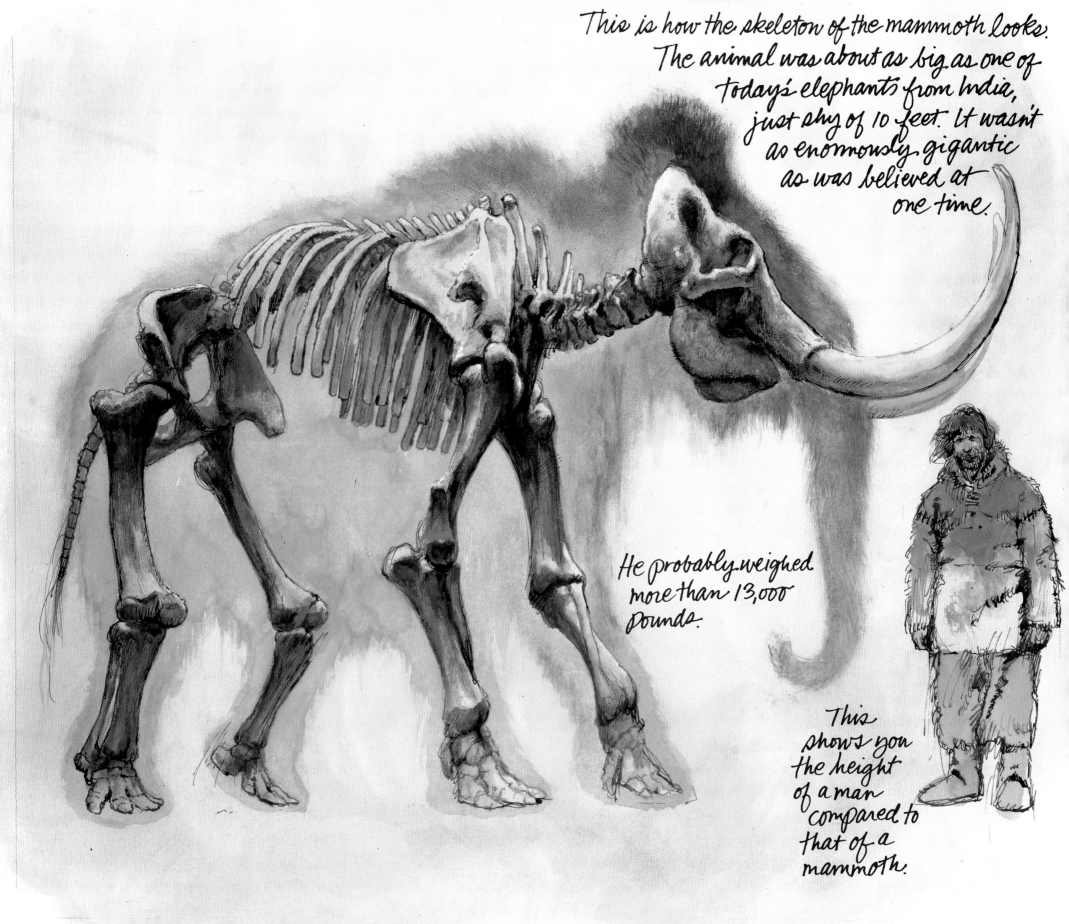

This is how the skeleton of the mammoth looks.
The animal was about as big as one of
today's elephants from India,
just shy of 10 feet. It wasn't
as enormously gigantic
as was believed at
one time.

He probably weighed
more than 13,000
pounds.

This
shows you
the height
of a man
compared to
that of a
mammoth.

213

The skeleton of a mammoth didn't hold any secrets for the people who lived at that time. They had handled every one of its bones.

In their language, the words must have sounded very different from the words we use today in our language — it is after all a long time ago — but all those bones definitely had their names, such as thighbones, tusk, shoulder blade, and mandible.

Our forefathers made good use of them.
Practically entire dwellings were built with
mammoth bones.
Two skulls were placed upside down
and two tusks were connected
to each other at the tips
forming the entrance.

The walls were made with stacked mandibles.

On top of these came shoulder blades and other bones.

The heavy hides were then pulled over the frame and, finally, reindeer antlers and things like that were hung and placed on top.

And then they moved in, the
mammoth hunters — that is, the
fathers and mothers and children,
some grandmothers and one or two
grandfathers; these were the people
who lived off the mammoths.

The whole world, cold as death and
silent as the grave, was theirs.

217

Inside, soup is being prepared,

probably mammoth soup.

Mammoth meat broiling on the spit

This is how you got warm water: in a pit lined with mammoth hide, snow was first melted with the aid of red-hot stones from the fire.

The "hot stones" shovel

A piece of mammoth hide

A humerus of the woolly mammoth. For fuel they often used fatty mammoth bones since there wasn't much wood here during the Ice Age.
It's all mammoth, and yet again mammoth.

Food, clothing, beds, fuel, building materials...
What would we have done without the
mammoths in those days?

A dwelling constructed of mammoth bones
has been excavated that was built with the
bones of at least 95 mammoths!

Just try to imagine that, 95 mammoths.
How do you ever manage to get hold of
such giants?

Not by staying inside, of course.
As the old saying goes,
that's how it is.

So let's go for it.

Always that howling, biting wind.

They knew exactly where the customary migratory paths of the mammoths were. But what could they do with their primitive clubs and spears against the hairy giants?

Little or nothing.

The only thing to do was to come up with some tricky ideas for trapping the animals, such as the pit. And then simply go to work with the biggest thing you have, a rather dull axe.

What else can you do?

You can drive into
a bog or onto thin ice.

222

And then, about 10,000 years ago,
it was all over.
The mammoths had had enough,
and the others as well. The woolly
rhinoceros, the lions and the
bears, the giant deer, all disappeared
from sight.

"On this earth they were not
seen again." They no longer
felt at home here.

As I sit
daydreaming on
my high seat I
like to remember them.
But, to be honest, when
suddenly a roe deer walks into
the clearing I hold my breath
in admiration and the mammoths
are completely forgotten.

_____

And I have more than enough
to keep me entertained!

Editor, English-language edition: Ellen Rosefsky
Calligrapher: Diane Lynch

Library of Congress Cataloging-in-Publication Data

Poortvliet, Rien.
    [Aanloop. English]
    Journey to the Ice Age: mammoths and other animals of the wild/
by Rien Poortvliet; translated from the Dutch by Karin H. Ford.
      p.   cm.
    ISBN 0-8109-3648-8
    1. Hunting—Netherlands—History—Pictorial works. 2. Mammals—
Netherlands—Pictorial works. 3. Mammals, Fossil—Netherlands—
Pictorial works. 4. Hunting, Prehistoric—Netherlands—Pictorial
works. I. Title.
SK223.N4P6613   1994
599.09492—dc20   94–1544

Originally published under the title *Aanloop*
Copyright © 1993 Uitgeverij J.H. Kok BV, Kampen, The Netherlands
English translation copyright © 1994 Harry N. Abrams, Inc.

Published in 1994 by Harry N. Abrams, Incorporated, New York
A Times Mirror Company
All rights reserved. No part of the contents of this book may be
reproduced without the written permission of the publisher

Printed and bound in Germany